Canines in the Wild

DINGOES IN THE WILD

Natalie Humphrey

Gareth Stevens PUBLISHING

Please visit our website, www.garethstevens.com. For a free color catalog of all our high-quality books, call toll free 1-800-542-2595 or fax 1-877-542-2596.

Cataloging-in-Publication Data
Names: Humphrey, Natalie.
Title: Dingoes in the wild / Natalie Humphrey.
Description: New York : Gareth Stevens Publishing, 2023. | Series: Canines in the wild | Includes glossary and index.
Identifiers: ISBN 9781538279595 (pbk.) | ISBN 9781538279618 (library bound) | ISBN 9781538279601 (6pack) | ISBN 9781538279625 (ebook)
Subjects: LCSH: Dingo–Juvenile literature. | Wild dogs–Juvenile literature.
Classification: LCC QL737.C22 H86 2023 | DDC 599.77'2–dc23

Portions of this work were originally authored by Elaine McKinnon and published as *Dingoes*. All new material this edition authored by Natalie Humphrey.

Published in 2023 by
Gareth Stevens Publishing
29 East 21st Street
New York, NY 10010

Editor: Natalie Humphrey
Designer: Rachel Rising

Photo credits: Cover Nicole Patience/Shutterstock.com; cover, p. 1 kwest/Shutterstock.com; cover, pp. 1, 3, 4, 6, 8, 10, 12, 14, 16, 18, 20, 22–24 Attitude/Shutterstock.com; cover, pp. 1, 3, 4, 6, 8, 10, 12, 14, 16, 18, 20, 22–24 donatas1205/Shutterstock.com; cover, pp. 1, 3, 4, 6, 8, 12, 14, 20, 22–24 Olga_C/Shutterstock.com; p. 5 Arun Sankaragal/Shutterstock.com; p. 7 Madeleine Steel/Shutterstock.com; p. 9 LKR Photography/Shutterstock.com; p. 10 Dominic Jeanmaire/Shutterstock.com; pp. 11, 21 Nicholas Toh/Shutterstock.com; p. 13 Nicole Patience/Shutterstock.com; p. 15 puyalroyo/Shutterstock.com; p. 16 jane mckinnon/Shutterstock.com; p. 17 FiledIMAGE/Shutterstock.com; p. 18 Susan Flashman/Shutterstock.com; p. 19 Nico Faramaz/Shutterstock.com.

Printed in the United States of America

CPSIA compliance information: Batch #CSGS23: For further information, contact Gareth Stevens, New York, New York, at 1-800-542-2595.

CONTENTS

Words in the glossary appear in **bold** type the first time they are used in the text.

IN THE AUSTRALIAN OUTBACK

For centuries, the dingo has roamed Australia from coast to coast. This cunning **canine** is an apex predator. This means the dingo eats other animals and is at the top of its food chain! The dingo is Australia's biggest **carnivore** on land.

What makes the dingo a master of the wild? Over the centuries, the dingo has **adapted** to fit its home. From the hot and dry Australian deserts to dense forests, dingoes can live almost anywhere.

Dingoes aren't originally from Australia, but they've lived there a long time.

WIDESPREAD WILD DOGS

Though dingoes are known as Australian wild dogs, similar wild canines live in Southeast Asia. The dhole is a wild dog native to Asia that looks a lot like the dingo.

In Australia, dingoes live in a few different **habitats**. They're found mostly in the western and central parts of Australia. Some live in hot, wet places. Dingoes can live in grasslands and forests, and even on islands. However, they're most famous for being animals of the Australian Outback, which is a large desert.

That's Wild!

Some scientists believe that dingoes **descended** from **domesticated** dogs. These dogs may have been brought over by Southeast Asian sailors and left behind.

To this day, exactly where and what dingoes descended from is not certain.

WHAT MAKES A DINGO A DINGO?

How do you know if you have spotted a dingo and not someone's pet? Look at its features!

Dingoes look a lot like domesticated dogs at first glance. They have thin, strong bodies that are perfect for running and for squeezing into tight spaces. They have pointed ears that always stand up and a long, narrow snout, or nose. Dingoes can grow to be about 4 feet (1.2 m) long, with a 13-inch (33 cm) tail. They usually weigh up to around 35 pounds (16 kg).

That's Wild!

The most common colors for a dingo are tan, gold, and red, but dingoes come in more colors! Their fur color helps them use **camouflage** in their habitat.

A dingo's teeth are longer than those of a pet dog.

LIFE IN A PACK

While some dingoes are solitary, or live on their own, most dingoes live in packs. Dingo packs are often made of a strong family unit. They usually include a mother and father dingo and their young. A pack may have up to 10 dingoes in it. Dingo packs have a **dominant** male and female dingo, also called the alpha male and female.

Dingoes communicate with each other through howling. They howl to let their pack know where they are and to let other dingoes know to stay away!

Unlike domestic dogs, dingoes don't bark.

PROTECTING THE PUPS

In a dingo pack, only the alpha male and female will **mate**. After nine weeks, female dingoes will give birth to around four to six pups, or baby dingoes. Before the pups are born, dingoes find a den to keep them safe in. Dens are usually hollow logs, rock shelters, or old rabbit warrens.

The pups will stay in the den until they're around two months old. During this time, the pups will be cared for by both of their parents and the rest of the pack. Dingo pups reach adulthood in about seven months.

That's Wild!

Ranks, or positions, in the dingo pack are very important. If another female besides the alpha has babies, the alpha female will often kill the other female's pups.

Dingoes can live around 10 years in the wild.

PERFECTLY ADAPTED

Dingoes have adapted perfectly to their Australian home. Their **adaptations** help them not only to live in many different habitats, but also to act as the top predator there.

Dingoes have great senses. Their pointed ears can move in many directions and give them excellent hearing. Dingoes have a great sense of smell, helping them find even the smallest **prey**. Their sharp eyesight helps them spot prey from a distance. When they need to, dingoes can also jump and climb trees.

That's Wild!

Dingoes can fit their body into any space where their head fits, helping them fit into small spaces to find water.

Dingoes can reach speeds of 37 miles (60 km) per hour. They can travel more than 25 miles (40 km) in one day.

WHAT'S FOR DINNER?

Dingoes are skilled hunters and can take down many different kinds of prey. On their own, dingoes usually hunt smaller animals such as rats, mice, lizards, and rabbits.

Packs of dingoes, however, can take down much larger prey. Working as a team, dingo packs can even hunt kangaroos! Hungry dingoes are also known to eat livestock, or farm animals, such as sheep and cows. Many people consider dingoes pests because of this, but most dingoes would rather find a wild meal!

When there isn't a lot of prey around, dingoes will **scavenge** for food or eat fruits and nuts.

HOME FOR A DINGO

Packs of dingoes will often stay in the same territory they were born in. Like many wild dogs, dingoes mark their territory with different scents.

A dingo's territory is often its hunting ground. Packs protect, or keep safe, the territory from other dingoes and travel only to find more prey. Dingoes will go great distances looking for enough food. They'll even hunt in the hot desert. Dingoes will travel between 6 and 7 miles (10 and 11 km) in one day to find food.

If a dingo wants to start its own pack, it leaves its old pack's territory to find its own space.

LIFE ALONGSIDE DINGOES

Even though dingoes have become symbols of the Australian Outback, they are still regularly in danger. Because they eat livestock, farmers view them as pests and often kill them. Farmers were so concerned about dingoes that they built one of the longest fences in the world to keep them out!

Another threat to dingoes is mating with other animals. Because dingoes belong to an animal group so close to dogs, they often mate with dogs. This has led to true dingoes being considered **vulnerable**.

That's Wild!

The dingo fence is 3,488 miles (5,613 km) long. It keeps dingoes from eating livestock. But without dingoes, populations of rabbits, kangaroos, emus, and non-native animals have taken over.

Dingoes help keep their **ecosystem** in balance.

GLOSSARY

adapt: To change easily to suit new conditions.

adaptation: A change in a type of animal that makes it better able to live in its surroundings.

camouflage: Colors or shapes on animals that allow them to blend with their surroundings.

canine: Of or relating to dogs.

carnivore: An animal that eats meat.

descended: To come from an animal of an earlier time

domesticated: Bred and raised for use by people.

dominant: The most powerful or strongest.

ecosystem: A natural community of life that includes living and nonliving features.

habitat: The natural place where an animal or plant lives.

mate: To come together to make babies.

prey: An animal another animal hunts for food.

scavenge: To search for and eat food left behind by people or other animals.

vulnerable: At risk of dying out.

FOR MORE INFORMATION

BOOKS

Downs, Kieran. *Dingo vs. Kangaroo.* Minneapolis, MN: Bellwether Media, 2022.

London, Martha. *Dingoes.* Minneapolis, MN: Abdo Publishing, 2021.

WEBSITES

DK Findout! Dingo

www.dkfindout.com/us/animals-and-nature/dogs/dingo/

Learn more about the dingo's amazing adaptations!

Fort Wayne Children's Zoo: Dingo

www.kidszoo.org/our-animals/dingo/

Learn more about dingoes from the people that help protect them.

INDEX